Time in Hell

The Battle for Buna on the Island of New Guinea

The story of Company F, 127th Infantry, 32nd Infantry Division Wisconsin National Guard, Sheboygan, Wisconsin

By Richard A. Stoelb

ISBN978-0-9818974-9-3

Sheboygan County Historical Research Center
518 Water Street
Sheboygan Falls, WI 53085

Printed in the United States of America

Dedication

It is with extreme pride and with deep heartfelt thanks that I dedicate this to all the men of the 32nd who "answered the call". But most of all I dedicate this to my dad, Roland A. Stoelb. This is their story about "time in hell".

Roland A. Stoelb
Camp Livingston, Louisiana 1941

In this publication quotations in italics are the words of Roland A. Stoelb.

Acknowledgments

To Beth, Katie, Mary, Kathy and Don at the Sheboygan County Historical Research Center. Thank you for "showing me the way".

Introduction

There are no extraordinary men, just extraordinary circumstances that ordinary men are forced to deal with.
Admiral William (Bull) Halsey

This is the story about the men of Company F, 127th Infantry, 32nd Division, Wisconsin National Guard who left Sheboygan, Wisconsin for Federal Service on October 20, 1940. They would fight in the Battle for Buna on the island of New Guinea against the Japanese in World War II.

My father, Roland Stoelb, served in Company F in the Pacific during World War II. But, he never talked about what happened over there. Since he didn't say anything about it I didn't bother to ask him any questions.

However, as I approached the time of entering the Army myself he started talking. Sometimes, as we sat together late at night watching the Tonight Show with Johnny Carson or an old movie on television, we would talk.

It seemed some event or remark would trigger a memory from him. It is difficult to describe, but sometimes it seemed more like he wasn't really talking to me. It seemed that for a moment he had gone somewhere else.

"We saw the Battle of the Coral Sea from the deck of our ship."

"We flew over the Owen Stanley Mountains."

"We didn't get any Thanksgiving or Christmas Dinner."

"I watched P-38's dog fighting with Jap Zeroes."

"When we were cut-off one of our guys went out and knocked out a Jap machine gun nest that had us pinned down."

"There's no glory in a foxhole."

"When the Japs opened fire even the guys that didn't believe in God prayed.."

"A Jap grenade killed a guy next to me and injured my right ear drum. My request for a Purple Heart was denied."

Roland Stoelb, in his chair at home outside of St. Anna, a place he sometimes talked about his war experiences; a place where it seemed that for a moment he had gone somewhere else.

After the bombing of the American fleet at Pearl Harbor on December 7, 1941 the Japanese rolled unchecked through the Pacific. By the middle of 1942 they were threatening Australia. They had to be stopped and that task would fall on the shoulders of the 32nd Division National Guard Units from Wisconsin and Michigan. Company F, 127th Infantry from Sheboygan, Wisconsin would find themselves right in the middle of it. This is the story of that little-known campaign as seen through the eyes of one soldier who was there.

As you read about the brutality that man inflicts upon man in war, do not pass judgment on these men. These were young men in their late teens to early 20s and they would be asked to do what no one at any age should be asked to do. War is not clean and sanitary. War is brutal, dirty, vicious, bloody, sadistic; *War is Hell.* War is not politically correct. It causes otherwise rational men to commit extremely irrational acts against others.

Today in war we talk about eliminating "collateral damage". But, there will always be "collateral damage" because damage is done to the ones who have to go in and do the dirty work because all else has failed. As you read this I ask you to please remember that. To truly understand one would have to have been there; been an actual participant in the action.

In war you trust one another; you look out for each other. You hope that you won't let the other guys down. It really is a matter of survival and you do what you have to do no matter how terrible, sickening or gut-wrenching that may be. Maybe that is why so many of them said little or nothing, or didn't write home about what they saw and did. The folks at home didn't need to know the "gruesome details", besides they already had enough to worry about.

The Forgotten War of the South Pacific

"The Marines, always the Marines. You would think they were the only ones there. They weren't the only ones fighting in the Pacific. What the hell do they think we were doing?"

From watching the movies and newsreels about World War II you get the impression that the war in the Pacific was fought solely by the Marines, and that they were the ones sent to fight the toughest and bloodiest battles. The war in the southwest Pacific, from New Guinea to the Philippines, was fought by the United States Army. These soldiers also served their "time in hell". For Company F, 127th Infantry, Sheboygan, Wisconsin, their time in hell came at "Bloody Buna" on the God-forsaken island of New Guinea. They could not forget.

The men of the 32nd were asked to do the impossible against a battle hardened, well supplied and fortified enemy. The 60mm mortar ammo they had was defective and frequently exploded in the mortar tube. The hand grenades they had left a trail of smoke during daylight and a trail of sparks in the dark giving away their positions. Japanese and Australian grenades didn't do that, so U.S. troops used captured Japanese grenades whenever possible and by "Midnight Requisition" helped themselves to Australian grenades. The one American 105mm Howitzer they did have was a very potent weapon against the heavily fortified Japanese bunkers; however, it was nearly impossible to move quickly and where needed in the tangle and mud of the jungle. The two available Australian heavy artillery pieces would sink into the mud when fired and required a couple of days to dig them out.

Furthermore, the troops were not adequately trained for combat of any kind, especially jungle warfare and survival in the jungle. They would enter combat and get their "on the job training" under the most horrifying of conditions. As bitter and costly as that training was, the survivors learned a lot about combat in the jungle, about Japanese defenses, the importance of sanitation, weapons maintenance, discipline and teamwork.

Their experiences taught lessons the Army had to learn to cope with the Japanese under conditions of tropical warfare. These men did do the impossible and they did it well.

In spite of the lack of training and the lack of equipment, this was first time in World War II the Japanese Army was decisively defeated. The men of the 32nd stopped the Japanese advance in the Pacific.

The 32nd Division achieved a number of important military firsts during World War II.

1. The first U. S. Division to embark for overseas service in one convoy after December 7, 1941.
2. The first U. S. Division to be airborne into combat over the Owen Stanley Mountains on New Guinea.
3. The first U. S. Division to fight an offensive action against the Japanese in the Southwest Pacific.
4. The first to supply itself by air.
5. Fought in six major engagements in four campaigns involving 654 days of combat or 15, 696 hours, more than any U.S. Division in any war. This was 48% of the total time the United States was in World War II. The soldiers spent forty-one months overseas, more than twenty-one months of them spent in combat. Their combat resulted in 35,000 Japanese deaths.
6. The first army of occupation in Japan.

My Dad

My Dad grew up during the Depression and Prohibition. He was born at 626½ Huron Avenue in Sheboygan, Wisconsin at 2:45a.m. the morning of February 29, 1920. The attending mid-wife was Mrs. J. Engelhard. His parents were Albert Peter Stoelb and Anna (Jaeckels) Stoelb. On March 14, 1920 he was baptized, Roland Albert Stoelb, at Holy Name Catholic Church in Sheboygan, Wisconsin by the Rev. D. F. Thill. Sponsors were Mary Stoelb and Frank Bersch.

He had six brothers, Bernard (Barney), Fredrich (Fritz), Jerome, Donald (Donny), Kenneth (Sonny), and Florian (Floyd), and four sisters, Mathilda (Tillie), Sidonia, Lucille and Marcella. Dad went to Pigeon River School in Sheboygan and completed the 8th grade. He never went to high school.

At the age of 14 he went to work on the 120 acre Bliss Farm and Dairy nearby He helped raise corn, oats and hay, and drove tractor and made repairs on machinery and milked 31 cows every day. For this he received "pay in kind", a gallon of milk and a pint of cream a day which he took home and gave to his mom.

From 1936 to October 1940 he worked for the John Hoogstra Coal Company in Sheboygan. He drove a four-ton coal truck delivering coal and wood making deliveries, servicing the truck and making repairs.

The family lived in a house located at the southwest corner of Eisner Avenue and Highway LS on the north side of Sheboygan. The PLENCO (Plastics Engineering Company) office now stands on the site.

"When I was 14 I got a job at Bliss Dairy. I didn't go to high school. A lot of the kids didn't. Any money we could make helped out at home during the Depression."

"In 1936 we were still in the Depression and jobs were still scarce and didn't pay much. When I was 16 I got a job driving a coal truck. I convinced 'Ol Man' Hoogstra I knew how to

operate a dump truck, which I didn't. I drove the truck out in the country and taught myself how to operate the dump levers."

"We heard the Sheboygan Guard Unit was looking for recruits to fill the ranks. A bunch of us thought joining the National Guard was a good way to earn some extra money. When you're 19 years old you sometimes a make spur of the moment decision that has far-reaching consequences. We went to drills one night a week and got paid $1.00 per drill. We got $1.00 per day when we went to summer camp at Camp Douglas. We drilled at the Eagle's Club on New York Avenue, later it became the Playdium."

Above: The new Eagles' Club, once located on New York Avenue between Seventh and Eighth Streets in Sheboygan is seen on May 21, 1927 at its grand opening. Three stories high with basement, the outer walls were constructed of buff faced brick and cast stone trim. The basement boasted twelve bowling alleys and a Rathskeller. W.C. Weeks was the architect.

Above: The main hall or auditorium in the Eagles' Aerie #277 was used for dances, games, conventions, exhibitions and drills by the local National Guard unit. It was the largest hall in the city at the time measuring 105 by 93 feet. The balcony was 18 feet wide and extended around three sides of the hall. Below: A 1950s image of the building when it was known as the Playdium. It was the home of the U.S. Air Force recruiting office, the bus depot. Next door to the west was Balde's Bar serving Kingsbury beer.

I started this "journey" to find out what my dad did during World War II. But, in addition to that I found something more important. I found my dad. I found out who he was and why he was the way he was. I learned to understand him.

Was he "perfect"? Of course he wasn't. He was human subject to all the "sins" that infect us all. I loved and respected him, but never feared him. He was a fair man, a compassionate man. He didn't even harbor any hate for the man who had killed his dad and his brother; he felt sorry for him having to carry that burden all his life.

I never heard him say a derogatory thing about the Japanese or any race. He was a very patient man. He was a proud man and proud of his family. It took a lot to get him angry. He never raised a hand to us or Mom. But, when he did get angry it was like he was more frustrated with you than really angry with you. He had seen enough anger, hate and death. He saw things no one should ever see. He had first-hand experience as to what unforeseen circumstances can force you to do, and how war changes you.

Most of his records were lost in a fire in 1973 at the National Personnel Records Center in St. Louis, Missouri. However, by matching his memories with official U. S. Army records, 32nd Infantry Division records, personal accounts from others he served with and various newspaper articles of that time, I was able to piece together what happened on New Guinea and the Battle for Buna from November 1942 to January 1943.

In September 1940 the war in Europe was a year old and all components of the United States Army intensified their training. On September 16, 1940 an advance echelon of the 41st Infantry began arriving at Fort Lewis, Washington. A short time later the entire 41st and other National Guard units were on post."

Ordered into Federal Service

(October 15, 1940)

"I was getting close to the end of my one year enlistment when we were called into active duty. In October the National Guard Units of Wisconsin and Michigan were ordered into Federal Service for one year. This was the first peacetime draft. Until we left for service we had to stay at the drill hall during the day but we could go home to sleep. We were under orders and had to get everything ready. We were scheduled to leave on a special train on Sunday the 20th."

The Sheboygan guard unit consisted of the Service Battery 120th Field Artillery, the Medical Department 107th Quartermaster Regiment, and the largest local guard unit, Company F 127th Infantry.

Train Departure Schedule

Train #1 left Camp Douglas at 12:10 p.m. Sunday for Milwaukee; left Milwaukee at 4:30 p.m.

Train #2 left La Crosse at 2:00 p.m. Sunday for Milwaukee; left Milwaukee at 9:30 p.m.

Train #3 formed at Oshkosh and Sheboygan; left Milwaukee at 11:15 p.m. Sunday.

Train #4 left Milwaukee at 11:15 p.m. Monday.

Train #5 left Eau Claire at 10:46 a.m. Tuesday.

Good-Bye Sheboygan

"The Service Battery left in convoy for training in Louisiana a few days before we did. We left on Sunday October 20th along with the Medical Detachment."

"Earlier that day we had a company picture taken in front of the Sheboygan Post Office. Late that afternoon we heard lots of people had gathered at the railroad station on Pennsylvania Avenue waiting for us. We marched from the Eagles Club to the railroad station. We didn't have much time to say good-bye because the train got there just a few minutes after we got there. We said our good-byes as we got on the train."

"My family was there. I had left home a few years before to live with one of my sisters due to a fight I had with my dad. But he came to the station to say good-bye and he shook my hand before I got on board. This was the last time we saw each other."

Above: Company photo of the young men from Company F of the 127th Infantry unit of the National Guard taken on Sunday, October 20, 1940 on the front steps of Sheboygan's Courthouse. Roland Stoelb is in the second row, fourth from right. This is just a partial image. *Sheboygan Press* photo.

...bers Of Company F, Largest Of Local National Guard Units, To Answer Call Of Their Country

The Sheboygan Press

OFFICIAL PAPER OF THE STATE OF WISCONSIN

SHEBOYGAN, WIS., MONDAY, OCTOBER 21, 1940

raft Lottery Set For October 29

irst Number
ill Be Drawn
By Roosevelt

Brooklyn Talk
Is Considered
By Roosevelt

Commanding Officers Of Three Guard Units

British Bomb Target
In Heart Of Berli

$39,152.05 Is
Fund Total At
Noon Today

Citizens And Workers Are Urged To Do Their Duty To Reach Goal Of $45,000 This Week

Nazi Planes
Raid Various
Objectives

Above: Headlines from the October 21, 1940 edition of the *Sheboygan Press* announcing the beginning of the draft.

Below: Soldiers from the local national guard unit Company F, 127th Infantry march along the railroad tracks to the train station in Sheboygan. Floyd LeMahieu is the standard bearer seen at right front. *Sheboygan Press* photo

Sheboygan bade farewell to the remaining units of its National Guard yesterday with hearty handshakes and good wishes, kisses and tears from wives, mothers and sweethearts. Above, the men are falling in as ordered by Captain Pelchen. Roland Stoelb at far left. Below: Soldiers lean from their train to kiss loved ones good-by. At far left, wearing the cap, is Erwin Heidenreiter. *Sheboygan Press* photos

The first day at Camp Beauregard for new recruits. Camp Beauregard, was located northeast of Pineville, Louisiana and was used as a World War II training area beginning in 1940. The camp, and several tens of thousands of acres of surrounding land, including Camps Claiborne, Livingston, Cook and Polk were used for the Louisiana Maneuvers, a training exercise involving almost 500,000 men, preparing them for the battles of World War II that they would soon be involved in. A full two-thirds of the U.S. military rotated through these war games.

"When we were all on board, the train headed south. In a few days we arrived at Camp Beauregard outside of Alexandria, Louisiana. It was hot and muggy there. At Beauregard we lived in tents with dirt floors and a stove for heat. Camp Beauregard was a place where you could stand in mud and have dust blowing in our face and there were bugs everywhere. "Camp Dis-Regard" in "Lousy-Anna" would be our 'home' for the next four months."

"In December, one of the guys, Ed Nelson (Sheboygan), from the 120th Field Artillery, went to town on a pass and was hit by a car. He died the next day."

Above: Company F, 127th Infantry photo taken at Camp Beauregard, Louisiana in 1940. The pyramidal-style tents identify this as Camp Beauregard. (Partial view)

Below: Camp Beauregard Company photo close-up. Roland Stoelb can be seen back row, third from left. The men are wearing their campaign hats. They could be crushed to fit into a duffel bag; they were typical Army hats, some too large, some too small, but worn with pride and individual style.

Sad Homecoming

"I came home on bereavement leave in December 1940 to attend a funeral. My father and my brothers Jerome and Florian (Floyd) were walking home early Sunday evening on December 14. They had gone to get some ice cream and were bringing it home and were almost there when a car crossed the centerline narrowly missing Florian, but striking Dad and Jerome from behind killing them both.

The driver admitted he and a buddy had consumed a few beers that afternoon. I found out later he was fined $50 and released because the judge agreed with his lawyer there was no evidence to warrant the charge of manslaughter against the driver.

After the funeral I returned to Camp Beauregard. The Red Cross had made arrangements for me to get home providing a round trip train ticket. They told me not to worry about anything. But, when pay day came at camp the price of the ticket had been taken out of my pay. The Salvation Army handed out cigarettes, candy bars, coffee and magazines to us for free. We could get the same stuff from the Red Cross too, but we had to pay for them. After that I always gave to the Salvation Army but not the Red Cross."

Camp Livingston

"We moved to Camp Livingston in February. It had just been built and was about ten miles northeast of Alexandria, Louisiana. It was better at Livingston. The tents had wood frames and wood floors. We had gas heaters, and we had a building to take showers in and to wash our clothes. Pay was good. We got fed, a place to sleep, clothing and medical care. Privates got $30/month. A Private First Class got $36/month, a Corporal, $54 and a Sergeant E-5 was paid $60 per month. It didn't pay to go to Alexandria on a pass. There were too many soldiers there and it was real easy to get into trouble without even looking for it. Local businesses had put signs in their windows that read, 'No Soldiers or Dogs Allowed'. *In New Orleans there was an amusement park near Lake Ponchatrain. They told us what places were 'off limits' in New Orleans and those would be where most of us headed for to have a good time."*

Camp Livingston, Louisiana, May 29, 1941, 127th Infantry, Second Battalion.

At right, Roland Stoelb on left with good buddy, Lowell *Kid* Lloyd at Camp Beauregard.

At left, Roland Stoelb, middle, examines a Pine Snake in Louisiana.

At right, Roland Stoelb posing with his Browning Automatic Rifle (BAR).

Selective Service Extension

In 1940 the 32nd Division, like all other National Guard units, if called into Federal Service, was made up mostly of men who could not be required to serve anywhere outside of the Western Hemisphere, and could not serve more than twelve months anywhere else.

"We all thought we would be home by October of '41. But, in August that year, less than four months before the attack on Pearl Harbor, Congress passed, by one vote in the House of Representatives, the Selective Service Extension Act *allowing the movement of all Army components overseas and extending our length of service for forever if necessary."*

Louisiana and Carolina Maneuvers

"During August and September we took part in the Louisiana Maneuvers, the biggest peacetime maneuvers in the history of the U. S. Army. Those maneuvers wiped out anyone who couldn't take the muggy summer heat. In October we were back at Camp Livingston. After maneuvers we got little rest. We were organized into a full regimental combat team to be added as a surprise force for the Carolina Maneuvers. In the fall we traveled by convoy from Louisiana to North and South Carolina"

This Changes Everything

December 7, 1941

The attack by the Japanese on the U. S. Fleet anchored in Pearl Harbor on the morning of December 7, 1941 brought the United States into World War II. Japan had joined the Rome-Berlin Axis alliance in 1940; thus, Germany, now an ally of Japan also declared war on the United States.

The United States and its allies decided that the Atlantic-European area would be the decisive theater of operations, and until Germany was defeated operations in the Pacific would have to be primarily defensive in nature. The basic reasons for that decision were:

1) The European Axis was the only one of our two separated enemies that could be attacked simultaneously by the three powerful members of the Allied nations of Russia, Great Britain and the United States.

2) The United States was the only member of the coalition free to choose which of its enemies to attack first.

3) The defeat of the European Axis would liberate British forces to apply against Japan.

After the attack on the U.S. Pacific fleet at Pearl Harbor the Japanese advance across the Western Pacific went unchecked. Japan coveted New Guinea. Once they controlled the island, they could isolate, and perhaps invade Australia. Even more important, possession of New Guinea would allow the Japanese to cut off the 8,000 mile supply line that ran from the west coast of the United States to Australia, via Hawaii and Fiji, thereby effectively ending our influence in the South Pacific. On the northern coast of Papua, New Guinea the Japanese wanted to set up a naval installation with an airfield at Buna Mission (called Buna Government Station in pre-war days).

The Japanese already had troops on the north coast of New Guinea and in mid-1942 tried twice to take Port Moresby, a key

Australian base in SW New Guinea. The first attempt by the Japanese was halted by the U.S. Pacific Fleet in the Battle of the Coral Sea (May 8, 1942). The second resulted in the Papuan Campaign (Papua New Guinea), one of the longest and most bitterly fought campaigns of the Pacific War.

"In late December 1941 we were assigned to "Force Magnet". These were the U. S. Forces stationed in Northern Ireland. We were scheduled to ship out from Fort Devens, Massachusetts to Northern Ireland for deployment to North Africa."

On February 17, 1942, General George C. Marshall ordered the transfer of the 41st Infantry Division, training at Fort Lewis, Washington, to Australia. Upon arrival the division's mission would be to protect Australian ports and air bases and provide garrisons for defense of its eastern and northeastern coasts.

Change of Orders

In March 1942 it was agreed the United States would take on the main responsibility for conducting military operations in the entire Pacific Area. General Douglas MacArthur was to command the Southwest Pacific Area.

"We were ordered from Fort Devens to San Francisco, California. We had no idea what was going on. All we knew was we were going west."

The main body of the 41st Division docked at Melbourne on April 6, 1942. And a few days later the 32nd Infantry Division and the rest of the 41st were put on orders for early departure to Australia.

"From April 10-14 we traveled on a Priority Troop Train from Fort Devens to San Francisco. When we got to San Francisco we stayed at Fort Ord. Some of the guys stayed at the Dog Track Pavilion, and the rest were "housed" at The Cow Palace. "

Before leaving for overseas the division was issued extra uniforms, new helmets, M-1 rifles to replace the old Springfields and new howitzers to replace the old WWI artillery.

On To Australia

"On April 19, 1942, we were packed into seven cruise ships from the Matson Line that had been converted into troop carriers. The Division had picked up 3,000 'selectees' along the way. They came from Wisconsin and Michigan as well as other parts of the country such as Illinois, Oklahoma, Wyoming, Louisiana, etc.

We got on board the Mount Vernon. Three days later we left San Francisco. The convoy was escorted by the cruiser U.S.S. Indianapolis and two smaller warships. As we left the docks we learned we were the first American division in World War II to be moved in a single convoy. We sailed past Alcatraz and as we went under the Golden Gate Bridge we knew we were really leaving home. A lot of us watched the coastline get smaller and smaller until gone. There was no turning back. We were at sea for three weeks and were only told that we were headed to Australia when we were well out to sea.

A lot of guys were not used to the rolling of the ship and made most of the trip topside "by rail". The smell of vomit was on all decks. The ships were packed full and we had to wait in long lines everywhere. We bedded down anywhere we could. The guys that slept below slept on sheets of plywood and as we got close to the Equator they came on deck to get away from the heat down below. Some of the guys wrote letters and some did reading. Others walked around the deck, did exercises, sang, or just watched the ocean. There were a lot of poker and craps games to help pass the time too.

We had classes on the Australian people and their customs and had abandon ship and fire drills. Crossing the Equator we were inducted into King Neptune's Court. One night, standing on deck, we could see flashes on the horizon. We learned later it was the Battle of the Coral Sea."

USS *Indianapolis*

USS *Indianapolis*, a 9,800-ton Portland Class heavy cruiser, was built at Camden, New Jersey. Commissioned in November 1932, she operated in the Atlantic and Pacific during the peacetime years. During the 1930s, she hosted President Franklin D. Roosevelt on several occasions, among them a voyage to South America in November and December 1936.

Following the U.S. entry into World War II, *Indianapolis* operated with carrier task forces in the southwestern Pacific until Spring 1942, when she took up station in the Alaska area. She served there for over a year, sinking a Japanese transport in February 1943. Later in 1943, *Indianapolis* became Fifth Fleet flagship. In that role, into mid-1944, she took part in operations to capture the Gilberts, Marshalls and Marianas, plus strikes on Japanese positions elsewhere in the central Pacific. She also participated in the Peleliu invasion in September 1944.

In February and March 1945, *Indianapolis*, again flagship for the Fifth Fleet, joined in attacks on Iwo Jima, the Japanese home islands and the Ryukus. During the latter operation, on 31 March 1945, she was damaged by a Kamikaze plane. In late July, following repairs, *Indianapolis* made a high speed transit from California to Tinian to deliver critical components for the first atomic bomb. She then sailed for the Philippines. Shortly after midnight on 30 July 1945 she was torpedoed by the Japanese submarine I-58 and sank quickly. Due to communications and other errors, her loss went unnoticed until survivors were seen from a passing aircraft on 2 August. Rescue efforts over several days saved only about a quarter of her nearly 1200-man crew.

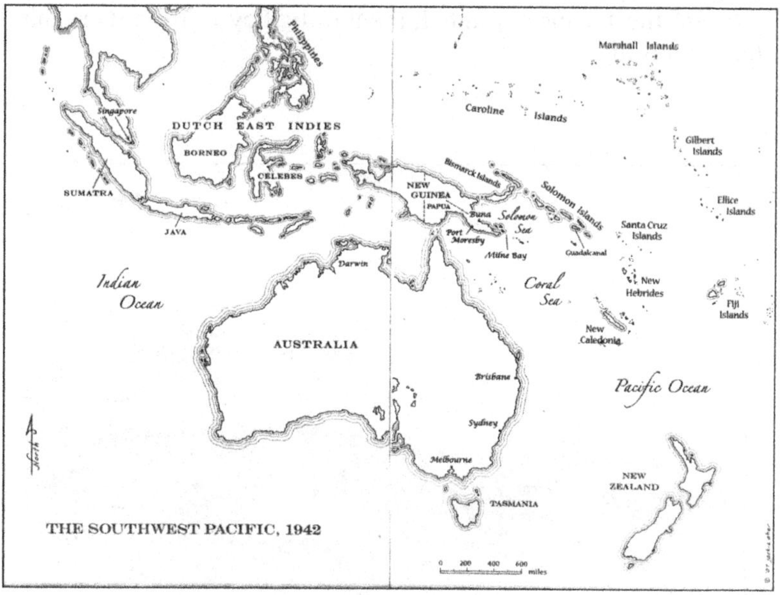

Above: A regional map of the southwest Pacific in 1942 showing Australia and the Island of Papua New Guinea located north of it.

Battle of the Coral Sea
May 8, 1942

Planes from the carriers *Lexington* and *Yorktown* attacked the Japanese task force on its way to Port Moresby, New Guinea. The Americans suffered heavier losses than the Japanese in the Battle of the Coral Sea; however, the fact that the Japanese turned back from Port Moresby left the strategic victory with the U.S. Fleet. The *Lexington* and *Yorktown* had been at sea when the Japanese attacked Pearl Harbor. These were the real "prizes" the Japanese wanted; thankfully they weren't there that fateful morning.

The Battle of the Coral Sea, fought in the waters southwest of the Solomon Islands and eastward from New Guinea, was the first of the Pacific War's six fights between opposing aircraft carrier forces. This is an image from the deck of the USS *Lexington*.

Arrival in Australia

"We finally got to Australia and docked in Adelaide on May 14, 1942. We went 8500 miles in 21 days. We headed for two camps. The 126th went to Camp Sandy Creek. The 127th and 128th went to camp Woodside 30 miles from Sandy Creek. When we got to Woodside it was dark. There was a blackout order and without lights we couldn't see anything. In the morning we got a good look and they were not what we had expected. The camp was just a bunch of tin shacks and warehouses with canvas roofs and no insulation. Our beds were burlap bags filled with straw. Later, we were moved to Camp Cable thirty miles from Brisbane. Camp Cable was better. The kangaroos were a problem; they would get excited and run into the sides of the tents. Brisbane was close to camp and a good place to go when you got a pass. While in camp "Lefty" (Hilarian) Hauch got spinal meningitis. "Lefty" and his brother Francis were in my wedding. "Lefty" was best man."

"We began toughening up in Southern Australia with lots of hikes and field exercises. We spent a lot of time in on the rifle range."

"Prostitution was legal in Australia and VD (venereal disease) was a big problem. Pro (prophylactic) Stations were set up in the cities and towns to try to control the problem. When you went to town you were supposed to pick up your prophylactics (condoms, also called "rubbers") at the station. They told us to buy a bottle of cheap whiskey too. The bottle of whiskey wasn't for drinking it was for pouring over your parts afterwards to disinfect them."

"The Army wasn't integrated then and the black soldiers were not assigned to combat units. They were assigned to the Quartermaster to distribute supplies or to the Engineers to build roads, bridges and such or to the mess. They tried to convince the Australian women that they were American Indians.

One time at Camp Cable the Regimental Chaplain called for an assembly of the regiment. He scolded us for our behavior when we were in Adelaide. He said there were over 700 pregnant Australian girls who identified the fathers as soldiers of our regiment. The chaplain said he was shocked at the possibility that a lot of soldiers may not have used their real name. One of the guys got up and yelled, 'Sorry, chaplain, we could have done better, but the officers keep us so busy we don't get to town very much.' Everyone broke out laughing and we were dismissed."

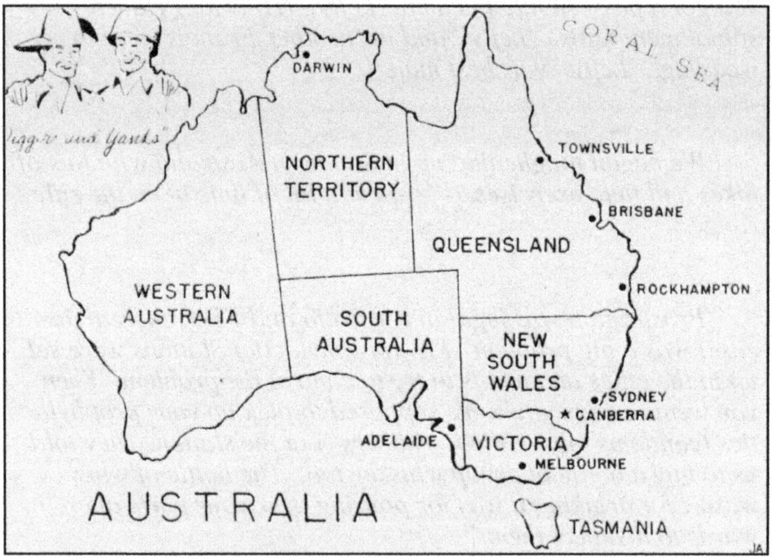

Provincial map of Australia.

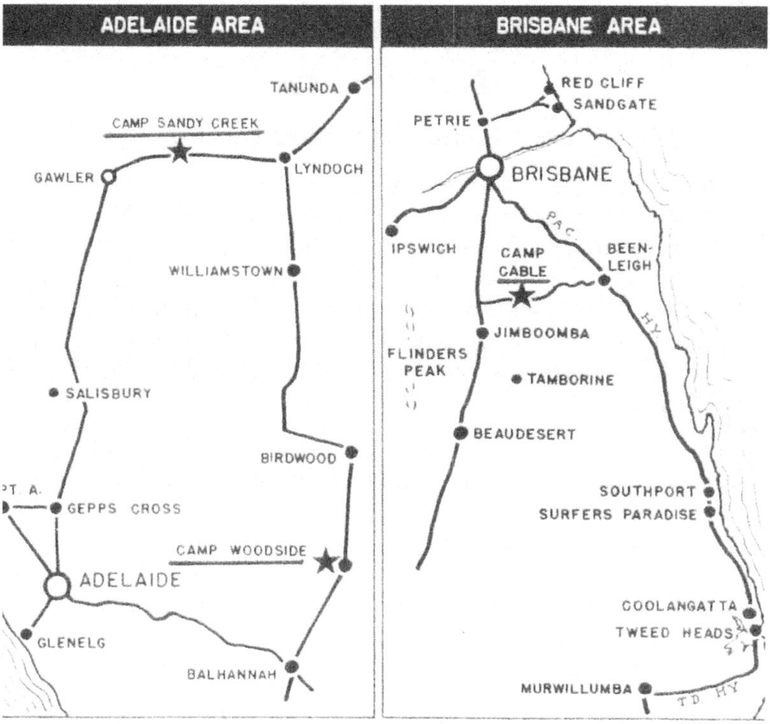

Above: Maps of Australian areas. Below: Map of the advance on the Buna-Gona area on New Guinea. Number 2 marks the final objective. Port Moresby is at the lower left.

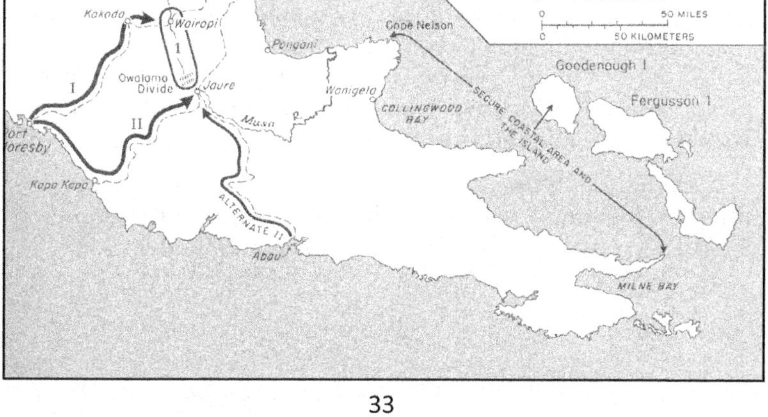

New Guinea

The Ultimate Nightmare Country

All American and Australian forces serving in New Guinea (Papua and Northeast New Guinea) were put under "New Guinea Force" with an expanded mission; it was no longer to be a defensive mission, but an offensive one. The Division's mission now was to halt further penetration of Australian New Guinea by the Japanese; to hold the crest of the Owen Stanley Range; retake Kokoda, the Buna-Gona Area and ultimately Lae and Salamaua.

New Guinea is a large tropical island located north of Australia covering an area of 342,000 square miles. It 1942 was a vast, primitive, almost unknown, largely unmapped wilderness of towering mountains, the Owen Stanley Range rising to 16,503 feet, and steaming tropical coastal jungles with stinking swamps. Papua lies at the southeast end of New Guinea and occupies an area of 90,540 square miles. At the time it was an Australian possession.

New Guinea is always hot, humid and wet with ever present mud that sucked at your feet and stuck to everything it touched. Daily temperatures averaged 95 degrees F and the humidity averaged 85%. It rained every day and every night. Tropical downpours of up to 10 inches a day were common. Rarely was there a dry place to sleep. Under such conditions tropical diseases flourished there---malaria, dengue fever, dysentery, athlete's foot, ringworm, beriberi, scrub typhus, diseases that make a man's scrotum swell to the size of a pumpkin. Every foot of travel through the jungle required extreme exertion. It was called the **"ultimate nightmare country"** and one American GI remarked, **"If I owned New Guinea and I owned hell I would live in hell and rent out New Guinea."**

Besides the diseases and the Japanese there were other things in the jungle that could kill you. Swarms of mosquitoes and flies constantly harassing you, chiggers, blood-sucking bats, rats as large as dogs, fleas, wild boars, poisonous snakes (Death Adder, Papuan Black, the Taipan), crocodiles, monitor lizards over six feet long and leeches everywhere.

The island was also called "Cannibal Island" because it was inhabited by primitive Aboriginal people, many of whom still practiced head hunting and cannibalism. The hair of these people is generally fuzzy rather than wooly giving Papua its name---"Papuwa" meaning frizzy or frizzled in the Malay language. The troops called them "Fuzzy-Wuzzies". It would be into this fearsome environment that raw Australian militiamen and American National Guardsmen of the 32nd Infantry Division would be thrown.

One military leader labeled the soldiers of the 32nd the "guinea pigs" of the South Pacific. They hadn't even been given the equipment that might have made life in New Guinea at least bearable. They had no specialized clothing. The fatigues they were issued were dyed for camouflage before leaving Australia. In the rain and extreme humidity the dye ran and clogged the cloth so no air could get through causing ulcers to erupt on their skin.

Simple items like machetes to cut undergrowth, insect repellents, waterproof pouches to keep salt, water purification tablets and quinine tablets from disintegrating in the rain, and cigarettes and matches from becoming sodden had not been issued. Most of the troops would suffer at one time or another from malaria, dysentery, dengue fever. More than 200 would succumb to the deadly scrub typhus.

The misery and terror of the jungle was so very alien to anything they had previously experienced. This unready division was out of necessity going overseas to fight a war the United States was not prepared for; a war against two great military powers and in the Southwest Pacific the Japanese held the upper hand.

The Papuan Campaign
The Buna Area

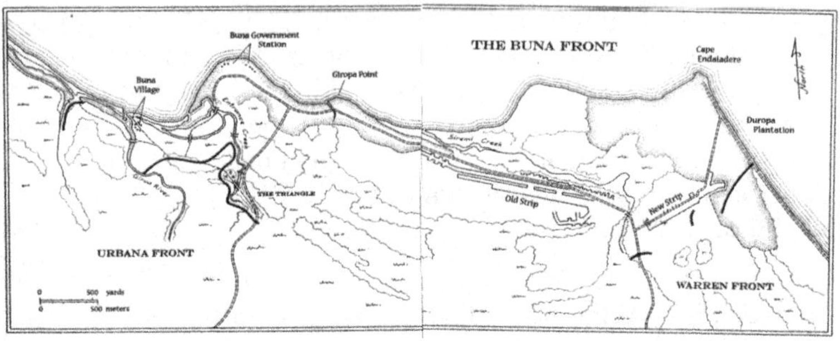

Above: The American forces are the Urbana Front and the Australian forces are the Warren Front.

Buna Village and Buna Mission were miserable little places located on the inhospitable northern coast of Papua New Guinea. Almost impenetrable growth covered most of the country with occasional patches of coconut palms. At the time the 32nd entered the area the rainy season had begun and the rivers and streams were in flood. Without machetes soldiers had to force their way through vines, creepers, brush, dense stands of razor-sharp Kunai grass and elephant grass higher than a basketball rim.

Formidable Japanese defenses covered Buna Village, Buna Mission and the Buna Strip areas. The main centers of enemy resistance were at Gairopa Point and Buna Mission. All approaches were very well covered by a series of heavily reinforced, well camouflaged bunkers of coconut logs and earth. From the bunkers machine guns could lay merciless cross fires upon approaching forces. The bunkers were also connected by a series of trenches allowing easy and almost undetected movement by the Japanese between them.

In the matted jungles these men would be forced to fight battles at point blank range; up close and personal. They would use anything that worked---grenades, fixed bayonets and sometimes their hands. An Australian historian called the fighting on New Guinea "The toughest fighting in the world."

An American historian conceded that the fighting there was "certainly the nastiest" of the war. Eric Bergerud described the struggle as "a knife fight out of the stone age" and Australian war correspondent, George Johnston, called it one of the "most merciless and most primitive battles."

The taking of the Buna area would give the Allies their first toehold in the area of Japanese defenses that guarded Rabaul and would deny the Japanese a base of operations on New Guinea at Port Moresby for invading Australia. The Japanese could not be allowed to go any further. The job of stopping the Japanese advance and evicting them from New Guinea fell on the 32nd Infantry Division National Guardsmen. Company F, 127th Infantry from Sheboygan, Wisconsin would be right there in the middle of it.

In September the 32nd and the 41st were assigned to Major General Robert L. Eichelberger, Commanding General I Corps, United States Army. General Eichelberger decided the 32nd Division would precede the 41st to New Guinea. The 127th and the rest of the 32nd prepared for immediate movement.

Chow in a foxhole.

The 126th Infantry (Ghost Mountain Boys) First American Infantry on New Guinea

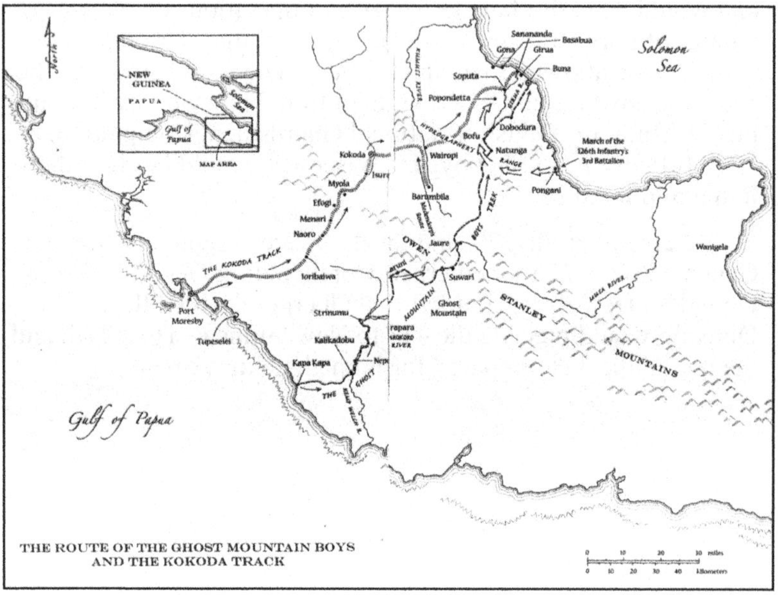

THE ROUTE OF THE GHOST MOUNTAIN BOYS AND THE KOKODA TRACK

General Eichelberger chose the 126th Infantry for immediate movement to New Guinea. The unit totaled 172 men in all, mostly from Grand Rapids, Michigan. They left by air from Amberley Field near Brisbane on September 15, 1942, for Port Moresby. The 126th arrived at Port Moresby that afternoon and were the first American Infantry to set foot on New Guinea. As they stepped out of the planes the first thing they noticed was the intense heat and humidity. The 126th (later named "The Ghost Mountain Boys" after a mountain on the Kapa Kapa Track) were to go over the Owen Stanley Mountains on the Kapa Kapa Track to engage the Japanese and to protect the right flank of the Australians on the Kokodo Track. This would deny the Japanese access to the Southern side of the island and access to Port Moresby. These Allied Forces along with the

Australian forces moving north along the coast of Papua from Milne Bay were to link up and converge on the Buna Coast.

The 127th Arrives In New Guinea

On November 14, 1942, the 127th and attached troops began

Above: Loading the ship at Brisbane, Australia for Port Moresby, New Guinea.

loading at Brisbane for movement by sea to Port Moresby, New Guinea. They would then proceed by air over the Owen Stanley Mountains to the Dobodura area. Dobodura was a grassy plain extending over several square miles and located fifteen miles south of Buna. It was an excellent site suitable for large scale air operations, even in the rainy season.

On November 16th the 127th left Port Moresby for New Guinea and the Japanese held positions at the Buna-Gona beachhead. The Urbana Force (Americans) was on the left, and the Warren Force (Australians) was on the right.

November 21, 1942: MacArthur to Harding; **"Take Buna today at all costs."**

Thanksgiving Day-November 26, 1942: By this time the Japanese were being driven back over the mountains by the 126th. Lack of food and disease was starting to take its toll on them. They were being hemmed in on the North Coast of New Guinea in the Buna-Gona Area. Sgt. William Fale, HQ, 127th, (Sheboygan) said to a war correspondent, "For a Thanksgiving dinner we had bombs served by the Japs.

"We got to Port Moresby on Thanksgiving Day. That night *we heard air raid sirens and then we could hear Jap planes overhead. We jumped into foxholes and then heard the falling bombs whistling. You could hear the 'wump-wump' of the bombs exploding near the airstrip."*

"Other units of the 32nd were already committed. Some were hiking over the Owen Stanley Mountains; some were going by boat along the coast. We were going to fly over the Owen Stanley's into combat."

"With full packs, weapons and ammo we got aboard the C-47's (DC-3's in civilian life). They didn't have any seats; we stood along the inside, and grabbed hold of a bar that was welded to the wall. The planes couldn't fly over the mountain peaks they were too high and they would need oxygen for us and the crew. So they flew between the passes and gorges banking left and right."

Above: Port Moresby, New Guinea as it looked when the headquarters of the U.S. Advanced Base in New Guinea was established there.

Below: Loading planes for transport at Port Moresby bound for the Buna area. *"With full packs, weapons and ammo we got aboard the C-47s"*

We Move To The Front

(December 6, 1942)

"We landed on a field near Buna and unloaded. We started immediately moving toward the Buna area. Walking up the trail toward the Japanese positions we could see the wounded lying there waiting for evacuation on the same planes that had just brought us in. Seeing all those wounded Australian and American GI's told us that this was no game; this was for real. It shook us up. General Eichelberger was there and told us as we walked by him that the Japs couldn't see or shoot straight, were starving and had poor weapons and equipment. The Japanese we came up against were tough and well dug in. They also told us the Japs were small. When I saw my first Jap Marine he didn't look small. He was over 6 feet tall."

"We moved up the track through fields of Kunai Grass over

"Walking up the trail . . .We could see the wounded lying there . . .".

6 feet high. That stuff had real sharp edges that could give you some real bad cuts. The trail went through a couple of native villages and the track was all water-logged going to the Buna Area. Going through the jungle was miserable. The heat and humidity didn't help any. The temperature was always 90 to 100 degrees and humidity to match. We carried about fifty pounds. We carried our weapons, ammunition, field pack, mess kit, entrenching tool for digging foxholes and a poncho. And the jungle had a rotten smell to it. It rained everyday and it came down heavy. It didn't take long to get soaked in rain and sweat. We were always wet and a lot of us got jungle rot in the crotch. We sweated a lot and our uniforms were white with salt. Eventually our fatigues and boots just started to fall apart. When our boots started to fall apart we looked for "better" ones on dead Japanese. It was hard to find a pair that fit. The Japs had small feet."

Walking through Simemi Village on the way to the front.

"It rained everyday . . . We were always wet."

"To get drinking water you dug a hole and waited for the hole to fill with water. The water table was only a couple of inches down. You filled your canteen and then added water purification tables, if they hadn't disintegrated, and hoped they would kill any germs. No matter how much water we drank we were always thirsty."

One time, during combat at Buna, the only ground water they could get was in a Japanese cemetery. No one ever forgot the taste of that water.

"We came up on a Jap soldier on the edge of the jungle. He was sitting in the sun but he had a rain poncho over him. It was real hot and muggy. We watched him for a while; but, he never moved. He just sat there. We approached him slowly. As we got close we figured out he was dead because we could smell him. One of the guys lifted the poncho off him and the smell really got bad. He was covered with maggots and flies were around him in hordes. Later in the fight for Buna we came across an American GI the Japanese had tortured and killed.

They had tied his legs to two trees that they bent over like spring poles and then cut the ropes."

"When I killed my first Japanese soldier I went through his clothing. I found his wallet and inside was a picture of a young woman holding a baby---I knelt there and cried. He had fought for what he believed in just like us, but he wasn't going home. There was no choice except him or me."

Buna Beach

The photo of three dead American soldiers lying in the sand on shoreline near half sunken landing craft on Buna Beach, Papua New Guinea is considered a war classic. Taken by George Strock of Life Magazine. It was not published until September 20, 1943.

This photo and other equally gruesome and graphic pictures of WWII were finally OK'd by the Office of War Information's censors, in part because President Roosevelt feared that the American public might be growing complacent about the war and its horrific toll. Even then, in the picture, the Americans' faces were not shown–a practice continued until Korean War to preserve soldiers' privacy in death.

At the time of the publication, these pictures shocked many readers. The *Washington Post* argued that the pictures "can help us to understand something of what has been sacrificed for the victories we have won." Images of dead soldiers appeared regularly after that.

The Situation

On December 7, 1942 at 6:00 a.m. the Japanese attacked the troops from Buna Village and Buna Mission. Machine gun and rapid rifle fire from the troops, reinforced by a fresh platoon from Company H, broke up the attack. At 1:35 p.m. after a 15 minute mortar and artillery barrage Company E and Company G started their attack. They were met by heavy opposition and fought off strong counterattacks from the village and mission. The attack made no progress. The next day they attacked again. At 2:00 p.m. artillery, mortars and machine guns opened up and at 2:15 p.m. troops started advancing. The results were the same as yesterday—no progress. The Japanese savagely contested every inch of ground. So far, twelve futile attacks had been made against the village costing heavy casualties and several days of effort.

On December 9th headquarters decided that the job of delivering the final attack on the Buna Village would fall on the fresh 3rd Battalion, 127th Infantry, which had just completed its movement to Dobodura.

In the early morning of December 14th troops of the 127th Infantry attacked Buna Village. At 10:00 a.m. upon entering the village they found the Japanese had gone; they had abandoned the village the night before. The Japanese left behind only a few guns, ragged clothing, canned goods and medical supplies. The thatched huts in the village had been blown to bits, the coconut palms splintered. It was a mass of wreckage.

The unopposed occupation of this mess was exaggerated by MacArthur's Headquarters into a major victory and HQ issued a communiqué to the New York Times of 15 December 1942; "Allies take Buna in New Guinea." However, hidden on the back page toward the end of the story is said "The main objective, Buna Mission was "still holding on".

Musita Island and The Triangle

Musita Island was located in a lagoon just to the south and west of Buna Mission. Seizure of the island would not only bring the mission under close-in fire but might supply a jumping off point for direct assault upon it from the south.

The Triangle was a trail junction where the Soputa-Buna track forks to Buna Village on one hand and to Buna Mission on the other was one of three good routes to the Japanese positions east of the river. This junction lies three quarters of a mile south of Buna Mission.

The Triangle would furnish an excellent line of departure for an advance through Government Gardens to the sea, a necessary preliminary to an attack on the mission from the southeast. On December 17th Buna Force issues orders for the capture of Musita Island and the Triangle. The island was to be taken on December 18th, the same day of attack on the Warren Front; and the Triangle the next day. An element of the 127th was assigned to take the island and the taking of the Triangle fell to what was left of the 126th. Col. Tomlinson gave orders to Company L, 127th Infantry, commanded by Capt. Roy F. Wentland (formerly Co. E, 127th, Manitowoc WI) was ordered to take the island the next morning.

It was December 18th and the all of the 127th Infantry would be available for these operations. The 3rd Battalion was already in line and had been for several days. After consolidating at Ango, the 2nd Battalion, which included Company F, had just begun moving to the front. Headquarters Company with Company G and H were also moving forward. The 1st Battalion was still being flown in and would come forward as soon as its air movement was completed. With the 127th steadily moving up Col. Tomlinson reshuffled the line. Company I, 127th took the place of 2nd Battalion 128th Infantry in the area between the island and the Coconut Grove. 2nd Battalion, 126th Infantry took over the Coconut Grove and moved troops into position above and below the Triangle. The 2nd Battalion, minus the mortar platoon of Company H, which remained behind, was ordered to Simemi for a well-earned rest. Company F, 127th Infantry reached the front and went into reserve along with Company E.

The south foot bridge linking the island to the mainland had been destroyed, and Entrance Creek could not be crossed, even at low tide.

The Division had no bridging equipment and the distance between the stream banks was too wide to fell trees for crossing, so swimmers dragged a cable across the stream. This worked and just before noon Capt. Roy F. Wentland, Co. L, 127th with two platoons and a light machine gun section of the company made it across.

The two platoons joined by a third moved along the eastern half of the island without opposition. As they started toward the bridge that connected the island with Buna Government Station (Buna Mission) they ran into heavy fire from concealed enemy positions.

In the firefight that followed, five men, including Capt. Wentland from Manitowoc, were killed, and six were wounded. Capt. Wentland, was then replaced by Lt. Marcelles P. Fahres (formerly of Co. F, 127th, Sheboygan). Heavy enemy fire continued giving the troops the impression that they were heavily outnumbered; thus, they pulled back to the mainland that night. The attack on the island had failed and the island remained in enemy hands.

Captain Meyer... you will cross... ...and that is an order."

Gen. Eichelberger decided to bypass the Triangle. Entrance Creek was unfordable and was spanned only by a footbridge that led directly into formidable bunkers in the Triangle, and attacks using the bridge had proved to be very costly. Therefore, the crossing for the Christmas Eve attack was to take place 400 yards downstream.

The men from Company K arrived at the crossing point mid-afternoon on December 21, and found the creek at this point was 50 yards wide. Wading into it they discovered the water was over their heads; moreover, enemy fire began to pour down on them. One man was almost blown out of the water by a grenade. The tide was coming in, water was lapping over the mangrove roots on the bank and the current was swift. Company Commander (Capt. Alfred E. Meyer; formerly Company F, Sheboygan), greatly perturbed at being ordered to make what he considered a suicide mission, pleaded with Col. Grose to let him wait until dark to make the crossing. Grose had already asked Eichelberger for more time but the general was "most impatient" and had refused. Grose told Capt. Meyer, **"Captain Meyer, you will cross your company here and now and that is an order."**

At 4:00 p.m. that afternoon Capt. Meyer sent troops into the creek to see if it could be forded. Not only could they not find a crossing point they were nearly "blown out of the water" by enemy fire from the other side.

The Crossing

Above: South footbridge over Entrance Creek to Musita Island. Below: Japanese-built bridge (north footbridge) from Musita Island to Buna Mission over Entrance Creek.

Capt. Meyer returned to his company and made several daylight attempts to cross, but enemy fire was too heavy. By nightfall a heavy rope was located and Lt. Edward M. Greene and several enlisted men began swimming to the opposite shore. Lt. Greene was killed almost instantly by enemy fire, and his body was swept away by the current. A few minutes later an enlisted man lost his hold on the rope and was swept away. Finally, one of the swimmers got the rope across the river, and the rest of that night (Dec 21/22) was spent getting the rest of the heavily weighted troops across in the face of continuing enemy fire.

By 2:00 a.m. on December 22, 47 men were on the other side; at daylight the total was 75. Company K suffered 54 casualties that night; 6 killed or drowned in the crossing, 8 killed and 40 wounded in the fighting at the bridgehead.

Early on the morning of December 22, Company K engaged the Japanese in a frontal attack in the bridgehead area. Company M's heavy weapons covered the area with fire from the west bank. Company I crossed on the footbridge and found a safe and easy approach to the bridgehead under the bank. Company I then went into position on Company K's right without a losing a single man. Bodies of troops who had drowned in the crossing the night of December 21/22 could be seen bobbing in the stream. Many of the bodies were not recovered. They were washed to sea by the current or consumed by the crocodiles that infested the creek. But the crossing had been made. There now was a strong bridgehead on the other side of the creek.

On December 22nd, having met no enemy resistance, engineers repaired the south bridge linking the mainland and Musita Island.

"A patrol went over a bridge and across the island without any trouble. When they got close to a bridge between the island and the mission they ran into heavy fire from the Japs. Two of our platoons and a machine gun section were called up front. By late morning the next day the Japs on the island had been taken care of. We were ordered back to the mainland."

In the evening, with the bridgehead and the northwest end of Government Gardens firmly secured, Col. Grose was given orders to attack across Government Gardens in an easterly direction the morning of December 24. Col. Grose had five companies of the 127th Infantry for the attack. Rifle units would start the assault supported by the heavy weapons crew of Company M located along Entrance Creek to the rear of the line of departure. Company G would be held in reserve and go into action upon orders from Col. Grose.

Below: A lugger or supply ship used to bring necessities to the Buna area on New Guinea.

Christmas Eve 1942

At dawn on Christmas Eve Company G crossed the creek. Company M's heavy weapons crews took up supporting positions along the bank of the stream. Company L replaced Company I on the left and Company I then extended to the right. Company K went into reserve. The two assault companies, I and L, each reinforced by weapons crews from Company M, moved into position along the line of departure.

At 6:00 a.m. artillery and mortars began firing and Company H on the island opened up on the mission with all of its weapons. At 6:15 a.m., covered by the rolling barrage, the troops jumped off along a 400 yard front. The Government Gardens were neglected and overgrown with shoulder high Kunai Grass. A narrow swamp had to be waded to reach the objective, a mile long stretch of coconut plantation on the coast between Buna Station and Giropa Point. As the two companies left the line of departure and began moving through the tall grass they were met by heavy fire and both were held up.

The fire was particularly heavy on Company I's front. The fighting was bitter and at close quarters. News came back to Col. Grose at his CP (Command Post) that Company I had suffered heavy casualties and had become disoriented. At 9:59 a.m. Col. Grose ordered Company I to the rear to reorganize. Company G was sent to replace Company I in the attack. Company G could not move ahead immediately because of intense enemy fire coming from hidden bunkers.

Company L was to have made the main penetration. They were given all available aid. Automatic weapons of Company M along Entrance Creek fired heavily in its support, as did the 81mm mortars massed south of the island. However, after making small initial gains Company L did not move forward. After seeing that the line had stalled Col. Grose ordered Company F, 127th to attack on Company L's right at 3:11PM. The result was the same. The line remained where it was and by evening it was no more than 150 yards from the line of departure.

"The attack had stalled and we were ordered in to reinforce the main unit. There was a lot of heavy fire coming from Jap

bunkers and pillboxes. We didn't have any better luck moving forward. Bob Keitel was killed that day."

Grose asked Gen. Eichelberger for more time to reorganize but the request was denied. Instead Col. Grose was ordered to resume the attack early Christmas morning.

Sometime upon entering action in the Buna Area, Bill Bommer (Detroit), a BAR man and friend of Floyd Lemahieu's, got separated from his unit and was caught alone by the Japanese. He killed eleven of them before they overpowered him. They tied him to a tree and used him for bayonet practice.

Humping ammo, artillery and mortar rounds.

Christmas Day 1942

Attack on the Government Gardens

Colonel Grose now had eight companies of the 127th Infantry at the front (A, C, F, G, I, K, L, and M).

"We were to attack across the Government Gardens. The plan of attack called for us and Company A to attack on the far left and push our way through to the coast. Two other companies were in the center of the line and were to push forward in their sector with us. Two other companies were to concentrate on getting rid of the Jap bunkers that covered the trail going through the gardens. One Company was held in reserve. We were to start our attack without any mortar or artillery cover. They told us there would be a diversionary attack from the island and one on Jap positions in the Triangle. The hope was the Japs would go for the diversion and leave their positions to meet the 'attacks' and when they did we would go in."

Christmas morning opened with a barrage of mortar and artillery fire on the mission. The guys on Musita Island made a great show of being ready to attack the mission from the island from across the north bridge.

"When the diversion coming from the island was at its peak, we were ordered to start our attack. Company A, was missing two platoons which had not yet arrived, was held up. Yesterday we couldn't move forward because of heavy resistance, but now the going was pretty easy. Sometime that afternoon we had made our way through the Gardens and had reached the coconut plantation."

Company F's position was now 200 yards west of the track junction, about 250 yards from the sea and about 600 yards from the Mission.

Isolated

"The Japs had been fooled by the diversion from the island. But, when they figured out what was going on they quickly regrouped and reoccupied their positions. And then it got bad, real bad. We got caught in cross fires coming from every direction. The Jap light machine gun was nicknamed 'woodpecker' because that's what it sounded like when it fired. Tat-tat-tat-tat

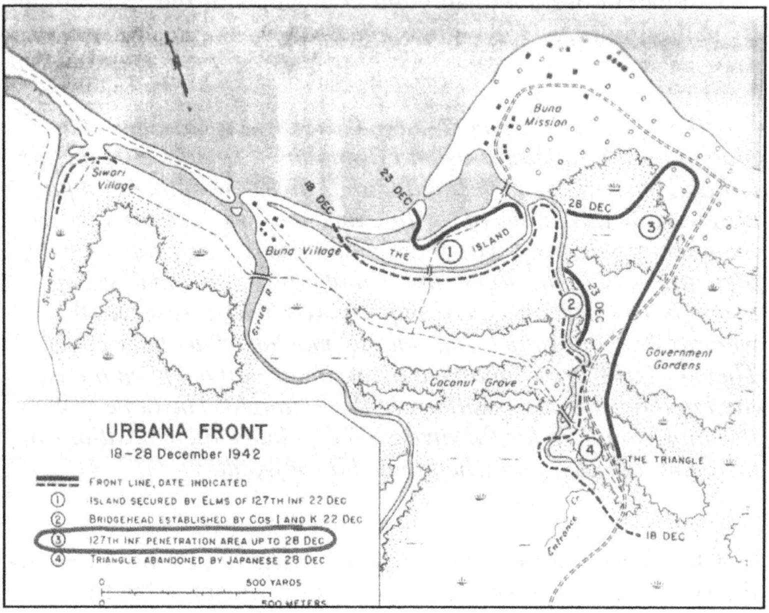

Above: A December 1942 map of the Urban Front showing the position of isolated Company F.

-tat. The Japs then counterattacked. Bullets whizzed through the air and shrapnel from grenades was flying all around. We beat off the attack but we took heavy losses; most of the 2nd Platoon had been killed or wounded. Steve (Engl) was wounded taking out a Jap machine gun nest that had us pinned down. Steve was in my wedding, too. We dug in and put up a defensive perimeter in a bunch of shell holes close to the plantation.

We were surrounded. We lost three Sheboygan guys that Christmas Day; John Hermann, Bernard Stegemeyer, and Les Vanderweele. Bob Keitel was killed the day before.

The Japanese kept up their attacks against us. In the after-

noon a detachment from Company A got through to us. Just as they were about to get to us the Japs had ambushed them and destroyed their weapons platoon. The rest of Company A reached the front later that day but couldn't reach us. There were big rats running over us at night, biting insects and leeches on everything. The radios got wet and quit working. There were a few open areas, but there wasn't much visibility in the jungle. You could be right in front of a bunker, or walk right past a Jap soldier and not know it until it was too late. The surrounding swamp area was loaded with snipers in bunkers and in the tree tops. It was almost impossible to move without being seen by them and getting fired at. They seemed to be all over the place. If you got caught in the open and got pinned down the sun would just bake you alive. They didn't always kill you but just wound you. No one came out to a save a dead guy but one or two would go out to save a wounded one and then get shot. The Japs even fired on medics trying to get to the wounded. Finally the medics just took the red crosses off or covered them with mud. If we located a sniper in a tree the 30 caliber machine gun was very effective against them."

Contact Lost

Since Christmas morning Urbana Force had not been able to contact the isolated troops their radios were wet and did not work. Three men from 127th HQ, ---Pvt. Gordon W. Eoff, PFC. William Balza and Sgt. William F. Fale (Sheboygan) volunteered for a patrol to protect linemen trying to get telephone wire forward. The patrol crept along a trail and located an enemy party. Sgt. Fale went ahead to investigate. He then proceeded to throw a grenade into the enemy group. Gunfire was then received from the enemy party and Sgt. Fale was seriously wounded.

One of the men in his patrol went out and removed him from the immediate area. Fale then offered to crawl out of the area rather than exposc the litter bearers to enemy fire. Sgt. Fale later died from his wounds received in this action. He was posthumously awarded the Distinguished Service Cross.

Sergeant Elmer D. Coon, of Black Creek, Wisconsin, facing heavy machine gun fire, made two attempts to cross a field to establish telephone communications with the isolated troops. Despite these heroic efforts all attempts to regain communications failed.

The next day (December 26th) Colonel Grose ordered Company C (less one platoon which was holding the area below the Triangle) in on the left. Their instructions were simple, "break through to the isolated troops near the coast, link up with them and form a corridor from Entrance Creek to the sea." On the far left the Japanese were still stubbornly resisting all efforts to dislodge them, and Company C made no progress that morning.

Major Edmund R. Schroeder (Company C, 127th, Oconto WI) ordered an element of Company C split into patrols and into the swamps north of the Gardens to deal with the enemy there to take Jap pressure off the isolated troops. The rest of Company C joined by Company B that afternoon, commanded by 1st Lt. John B. Lewis, continued the attack frontally.

The Japanese made good use of the terrain. Advances through narrow passages in the swamp and jungle had to penetrate their strongest defenses. Thus, the attack made little progress and it was becoming apparent that Company C was not going to break through to the isolated troops.

Colonel Grose ordered the Chief of Staff, 32nd Division, Colonel Bradley, to go out to the isolated company and to bring back a report on their condition. At Grose's request, Bradley, also acting executive officer of the 127th Infantry, was to be accompanied by Major Schroeder and 1st Lt. Robert P. McCampbell, (from Nebraska) S-2 of the 2nd Battalion, and a platoon of Company C, led by 1st Lt. Ted C. Johnson. The patrol left immediately carrying wire, ammunition and food.

"We were in a swamp and in pretty bad shape. We couldn't take care of the wounded guys and couldn't bury the dead. It wasn't good but we held our position. A patrol reached us late afternoon the next day. They brought ammo and food. A small patrol returned to HQ. A Major Schroeder and a Lt. McCampbell stayed with us. Then we waited for reinforcements."

Later that evening, Colonel Bradley, accompanied by a small patrol, returned from the perimeter with a full report to Colonel Grose.

"On the 27th we were still isolated and what was left of us expected the worst. We were dug in the swamp laying in water, mud and muck. Looking up we saw a bunch of Jap Bombers and Zeroes flying low over our position. You could tell a Jap Zero because it sounded like an old washing machine. It looked like they were about to attack us. All of a sudden, four P-38 Lighting's came out of the sun taking out the bombers and "dog-fighting" with the Zeroes. Those planes saved us that day."

One of the P-38 pilots was Richard Bong from Poplar, Wisconsin. He went on to become an ACE, but his first kills were a Jap Bomber and a Jap Fighter over New Guinea on December 27, 1942.

Company L was ordered out of the line to rest and clean up. The attack was resumed early and Col. Bradley, on Grose's orders, was in command of Companies B, C and the detachment of Company A. Company C and the detachment of Capt. Harger's company were held up by enemy bunkers north of the Gardens, but Company B made good progress all day. Japanese Zeros bombed the Urbana Front in the morning and in the afternoon the Japanese set a fire in the Gardens that swept across the entire front; however, this only slightly disrupted Company B's advance.

At 5:00 p.m. Company B moved through the Gardens to the isolated troops. Bradley, who had been at the front all day, reported the troops on the Urbana Front were suffering from battle shock and unable to advance. One night many years later while watching a movie on TV a scene depicted the time Gen. George S. Patton was visiting wounded soldiers in a hospital in Italy. He came upon a soldier suffering from "battle fatigue". Patton slapped him and called him a coward. My Dad said "*That Son of a Bitch didn't know what the hell he was talking about.*" Now I know why he said that.

December 28, 1942

"Maj. Schroeder had the forward perimeter well organized for action. The next day a trench was finished across an open area to us so the wounded could be evacuated and they could bring us ammo, food, and water. We had a working telephone, ammunition and food had been brought forward. Once the wounded were evacuated we were organized for attack. After four days Company C on our left and Company G on our right broke through to us."

The rest of Company A moved in soon after and a broad corridor from Entrance Creek to the Coconut Plantation had now been established.

On December 28, 1942 *Time* magazine said this about the battle for Buna; "Nowhere in the world today are American Soldiers engaged in fighting so desperate, merciless, so bitter, or so bloody."

An examination of the Triangle revealed 14 bunkers and innumerable trenches surrounded by swamp making it quite evident how the Japanese could hold off attacks for so long. Urbana Force was now able to use the trail through Government Gardens as far as the Coconut Grove, and it also now controlled the track junction along the coast. A forward element of Company B was only 120 yards from the sea. The corridor from Entrance Creek to the coast had finally been established. Buna Mission, now cut-off from Giropa, now lay open to assault.

"We were deep in the Coconut Plantation and Floyd LeMahieu was First Sergeant. He had just crawled back through the trench to rejoin the company after giving his "morning report". Woody (Sgt. Duke) asked him for a can of c-rations. Floyd bent down and a Jap sniper opened fired on him. He dove for cover but a bullet hit him in the left foot. Lowell Lloyd, from Sheboygan Falls nicknamed 'Kid', called to him to see if he was OK and he was. Then Kid told us all to sit still. It was real

quiet and then we heard a burst from his Thompson submachine gun and then quiet again. Then we heard a real long burst. When Kid came back we asked him; 'Did you miss him the first time?' 'No, I got him.' He told us he used an entire drum of ammo in the second burst to cut off the top of the coconut tree where that sniper was hiding."

Lowell 'Kid' Lloyd

"Maj. Schroeder was in command and we had been attacking toward the mission. We didn't make any progress. The Japs were holding on along the coast. We had a secure position facing the mission and we were holding the line along with three other companies."

Elements of two others companies were in place in the Gardens on both sides of the corridor and a couple more were in the center of the corridor and one was to the east of it facing Giropa Point.

"When we moved against the mission the Air Force made bombing and strafing runs over it. They didn't bother to find out where we were and where the Japs were. You feel pretty helpless being bombed and strafed by your own guys. We were so pissed that we shot back at them."

When the soldiers were back in Australia Steve Engl would check the bars with his squad looking for pilots that bombed and strafed us. When they found them they would beat them up and cut out the part of their uniform with their ribbons. The next morning the "trophies" would be found pinned on the company bulletin board.

Final Assault

On New Year's Day, January 1st, 1943, Urbana Forces launched what it hoped would be the final assault on Buna Mission.

Early in the morning Company B attacked eastward toward Giropa Point while artillery and mortars laid down a heavy barrage on the Mission. The rest of Urbana Force struck at the Japanese line around the Mission. Companies A, L, M, and Company F moved forward with Companies I and D immediately to the rear and made some progress.

Firing a 60mm mortar on Buna Mission

New Year's Day 1943

"Bud Norgaard and Warren Schneider were killed on

Above: 50 Calibre Browning machine guns fire on an enemy position.

Below: 37mm anti-tank gun. Canister rounds fired from this gun were very effective against the Japanese in the jungle.

Above: 32nd Infantrymen firing into a dugout used by Japanese to fire upon Australian troops transporting ammunition.

Below: Soldier looking into the doorway of Japanese pillbox, December 23, 1942.

New Year's Day. Bud's name was added to the American Legion Post in Sheboygan Falls; the Frederick-Mueller-Norgaard Post. That evening an outpost reported seeing Japanese troops trying to swim from the Mission."

At 2:00 a.m., Saturday, January 2nd, twenty enemy soldiers with heavy packs were spotted making a break for the beached landing barges on a sand pit west of the Mission. The troops let them pass and then with rifles, machine guns and hand grenades killed them all. It was discovered the dead Japanese had three machine guns with them, and their packs contained food, medicine and personal effects. This was evidence sign they were trying to make an escape.

"At daylight we saw large numbers of Japanese in the water trying to swim to safety. Some were in small boats and some were swimming or hanging on to anything that floated. They were easy targets. Artillery and machine guns opened up on them as soon as they were spotted. Riflemen and machine gunners from the village area also fired at them. P-40's came over and strafed the Japs in the water."

Artillery pounded Buna Mission. Mortar shells were exploding throwing up clouds of dust. White phosphorus shells set grass, trees and wooden buildings on fire. The barrage didn't weaken the resolve of the Japanese; they remained a very determined adversary still full of fight with no desire to surrender. To the Japanese soldier it was better to die for the Emperor with honor that to surrender and bring disgrace upon yourself, your family and the Emperor. Although the Buna Mission was partially evacuated there were still enough Japanese left there and along the approaches to give Urbana Force a hell of a fight.

At 10:15 a.m. the American troops moved forward from the spit, the swamp and us on the east.

"We moved out, the Japanese opened fire on us from the bunkers, trenches and trees. Maj. Schroeder was in a forward observation post at the time when a sniper shot hit him in the

head and killed him."

At first the attack went smoothly. Several points in the mission area were now on fire from the phosphorus shells. The fires exposed a whole line of enemy bunkers to Allied fire. Japanese attempting to leave the exposed bunkers were cut down by machine gun fire. Several of the huts caught fire as shells exploded in the trees. When they tried to escape from dugouts beneath the burning huts they were greeted by a hail of gun fire and most of them were killed. The remaining holdouts continued a last-ditch effort of resistance and had to be rooted out bunker by bunker and dugout by dugout by grenade, machine gun and submachine gun fire.

"The "Fuzzy-Wuzzies" didn't like the Japanese because they destroyed their gardens, took their pigs and raped their women. They liked us and we used them as porters for our supplies and equipment and as litter bearers for our wounded. We paid them in cigarettes and c-rations. They really liked American cigarettes. They would bring in Jap heads and want one cigarette for one Jap head. After awhile we figured out that they were bringing in heads from Japs that had been killed in battle. We told them we would not pay for "cold" Jap heads. They would only get paid for "warm" Jap heads. Another time we loaded some Japanese prisoners on to a C-47 for transport back to Moresby. We couldn't spare any of our guys to guard then so we used "Fuzzy-Wuzzies" as guards. We learned later that when the plane landed there were no Japs on board. When asked what happened to them one 'guard' said, 'We fly over mountains. Japs get scared jump out.'"

Mission Compound Entered and Secured

"Late afternoon we broke into Buna Mission compound. There was complete destruction. Jap bodies were everywhere. The buildings were on fire, tops blown off trees and the ground was covered with shell holes. 'Maggot Beach' was covered with abandoned weapons and destroyed landing barges. The smell of dead and burned bodies is unforgettable. We flushed out any remaining Japanese from their hiding spots and killed

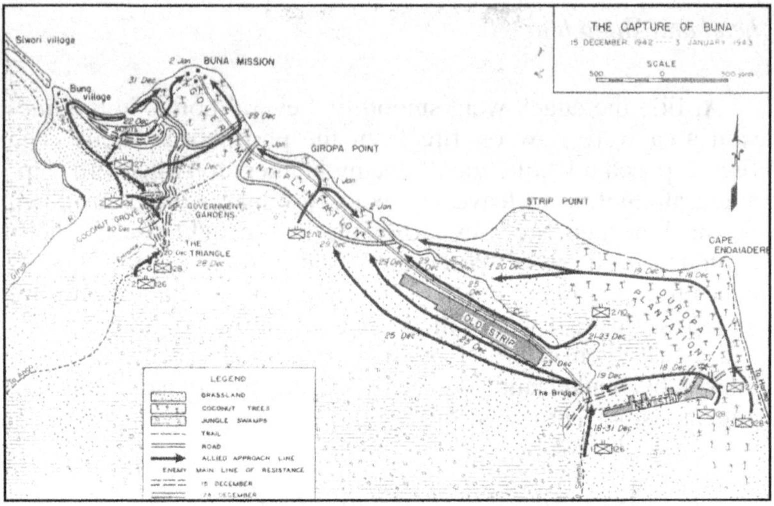

The Capture of Buna, December 1942- January 1943

them or we just buried them where they hid."

"Lester Holst was killed sometime in December in the fighting for Buna."

By 5:00 p.m. the fighting was over although it would be several days before the last Japanese would be eliminated. After more than six weeks of fighting, the entire Buna area was finally in Allied hands.

An "on and off" rain fell all day on Sunday, January 3rd. Tillman Durdin of the New York Times observed that the front from Buna Village to Cape Endaiadere was "almost peaceful". A shot could be heard now and then and an occasional burst of mortar fire, but for the most part the only sound was the rain falling on the shattered palms. Some of the troops went swimming in the ocean; some washed their fatigues (for the first time in weeks). And some "simply slept the deep sleep of exhaustion, curled up under shell-shattered palms or in sandy foxholes." Durdin also noted that their faces were gaunt and haggard, many were flushed with fever; and that the elbows and knees poking through their ragged uniforms were pathetically bony. However, their rest was just a brief respite. Another three weeks of battle still lay ahead of them before the "Victory at Buna" would be complete.

Mopping Up

"For days after taking the Mission we searched the area eliminating the last of Jap resistance. We searched the coconut groves all around the sector picking off snipers in the trees. Before going into an area we would spray the tree tops with machine gun fire to clear any of them out. We were told that 'if they don't stink stick em' to see if they were 'playing possum'. We made sure they didn't have to 'play possum' anymore. Some had committed suicide. They stuck the barrel of their rifle in their mouth and pulled the trigger with their big toe. Because we had cut off all supplies to the Japanese they were in bad shape and had turned to cannibalism. They had sliced the 'hams' off of dead GI's, Australian soldiers and their own dead to cook and eat."

Troops of the 32nd examine booty after taking Buna Mission. *"Pistols, swords, flags and personal affects were some of the souvenirs taken as trophies."*

Above: Coconut log bunker with fire trench entrance in the Buna Village area. The trenches allowed the Japanese to move from bunker to bunker unseen. Below: Interior of coconut log bunker reinforced with sand-filled oil drums.

The last resistance to Buna Mission was quelled by noon. In a little thicket on the beach a few Japanese held out in bunkers. Routed from the bunkers they took cover behind a wrecked barge on the beach and continued to fire. They were finally killed by a high explosive charge that blew both the barge and the Japanese to bits. The last enemy machine gun nest was eliminated by late afternoon.

"Burial squads buried the bodies of our dead and Graves Registration recorded their locations. The men who died there were buried on the battlefield with a rifle marking each grave. They were about 150 dead Japanese in the Mission area. We buried the Japs where they fell and some in mass in their trenches. Pistols, swords, flags and personal affects were some of the souvenirs that were taken as trophies. One GI added to his lucky necklace of gold Jap teeth."

By January 4th, the only Japanese left in the area from Buna Village to Cape Endaiadere were roving groups and individuals hiding out in the jungle and swamp by day, and coming out at night to prowl through the American-Australian lines for something to eat. Some 190 Japanese were finally buried at Buna Mission and another 300 at Giropa Point. Fifty prisoners were taken; Warren Force took twenty-one horribly emaciated Koreans and one Japanese soldier; Urbana Force took twenty-eight prisoners, mainly Chinese and Koreans. The few Japanese among them were captured when they were caught naked and unarmed as they swam in the sea. An Allied communiqué proclaimed there was "A striking victory" at Buna. The enemy has been "annihilated".

General Eichelberger, Commanding General of the American forces in this area issued the following memorandum to his men: "Immediate operations in the Buna area have come to an end and a hard-fought campaign has resulted in victory. The enemy with veteran troops was entrenched with his rear on the ocean, and his left flank on the ocean and his right flank resting on unfordable tidal streams. Across most of his front were deep swamps. American troops in this area have had their baptism of fire. Brave men have been killed or wounded that victory might be attained. You are now veterans and the lessons which have been learned in this campaign must serve to reduce our losses in the future and bring further victory. My congratulations on the victory you have won."

It's Not Over Yet

In his Order of the Day on January 3rd, General Herring told the troops the battle for Buna was "but a step on the way." The difficult job of cleaning out the enemy in the Sananda area would "not be any easier than Buna." The 127th was in position near Tarakena and the 18th Brigade, the tanks, and most of the guns in use at Buna were now available for use on the other side of the river. The time to move

Buna Mission after being leveled by 32nd Division mortar fire. January 1943.

Above: Japanese dead on the beach near Buna Mission. They were killed trying to escape by sea. January 3, 1943.
Below: Emaciated and beaten Japanese prisoners, many suffering from dysentery and starvation, being taken to the rear for interrogation.

on the enemy's Sananand-Giruwa position had come. This wasn't over yet.

Although the 32nd Division had some beach-defense responsibilities in the Buna area, only the 127th Infantry was now actively engaged in offensive action.

Col. Grose felt prompt action was necessary to meet the threat of a Japanese attack from the west and retrieve the lost bridgehead at Siwori Creek.

On the night of January 4th artillery fire was laid down. By early morning on the 5th Company G, 127th Infantry commanded by Lt. McCampbell crossed Siwori Creek. Company F followed shortly under command of 1st Lt. James T. Coker. The crossing was slow. They only had two small boats to ferry men and supplies across. Meeting no opposition the crossing was completed by 9:00a.m.. The troops quickly regrouped on the opposite bank and began moving westward toward Tarakena Village. Company G moved along a narrow, exposed coastal track; Company F, in the difficult swamp area, covered it on the left. The rest of the regiment was also in position to move on the village. Company E joined in and the advance continued against considerable enemy resistance.

On January 5 and 6 the Japanese made several strong stands retreating only at the point of their being overrun.

On January 7 the troops were within 500 yards of the village; again the Japanese made a stand. Company E moved its 37mm gun onto a sand spit, and with canister rounds brought enfilading fire upon the enemy. With this support Company F with only 72 men left, along with Company G pushed the Japanese back and captured 5 machine guns that evening. By 6:30 p.m. on January 8 the village was entered and by 9:30 p.m. the fighting was over. General Eichelberger reported the next day to MacArthur, moral was good and he heard some men singing for the first time.

The End

The 127th had to fight its way both south and west through "a swamp that stinks like hell." Even though the weather had somehow changed for the better the advance at first was slow.

But the attackers pressed the Japanese hard and resistance crumbled. The village of Giruwa was taken almost without opposition on January 21, 1943.

On January 22 some mopping up remained, but the 32nd Division's active operations in Papua were at an end. Guadalcanal was not declared won until February 9, 1943.

Due to higher priority demands placed upon aircraft and ships, it would not be until April 1943 that the 32nd would return to Australia.

"We flew to Port Moresby and then went by boat to Brisbane. We stayed in an R & R area on the coast for a few days. The first night back most of us went to town and got drunk. We were sent back to Camp Cable to get new clothing, receive replacements, get medical care and begin training for the next campaign. I was assigned to light duty as a clerk for nine months and then was transferred to the Quartermaster as a truck driver."

Of these two earliest campaigns in the Pacific, each lasting six months almost to the day, Buna had been the bloodier. A heavy price was paid for the all-out victory at Buna. From their bunkers on the New Guinea coast the Japanese between mid-November and January 22 cost the Allies 6,419 ground troops killed, wounded and missing, a large proportion of those battle casualties (beginning 22 July) of 8,546.

1944

The Northern New Guinea Campaign

Aitape, New Guinea

Taking Aitape would provide supply and air bases in the region of great strategic importance while denying them to the enemy. Land-based fighters could provide air cover over Hollandia from Aitape. And having ground forces in the area would also cover Hollandia assault forces from any movement by the Japanese troops against them.

"We were sent to northern New Guinea to take an enemy airfield near Aitape as part of "Persecution Task Force. On April 22 we moved out from Finschlafen and made a beach landing. The troops got into small landing craft and headed for a small island called Celio. They hit the beach and ran for cover. Almost as soon as they hit the beach they began running back to the water. They were covered with fire ants and ran into the water to get rid of them. They met no enemy resistance on the island and they set fire to the native houses as they moved."

"For a few days it went well. The beach was full of supplies and we had tents on the beach. We patrolled the coast and set up outposts all along. We moved further into the jungle to the Drinimore River. Some Japanese units were trying to join other units up the coast. We ran into heavy resistance and we heard there was a big Jap force in the area. We moved out of the swamp to higher ground and they told us to dig in. About three in the morning we heard a bugle and the Japanese launched a Banzai attack. Our artillery opened up and barrage after barrage just pounded the area. They Japs were being cut down by shrapnel and flying rocks and boulders. Others were just blown to bits but they just kept coming. When they got to the barbed wire the machine guns opened up and the Jap bodies piled up on the wire. We kept firing until we ran out of ammo. The anti aircraft guns fired air bursts above them. No matter what we threw at them they just kept coming. It didn't last long, maybe about half an hour or so and some Japs did get through. The next day we went in and saw everything was blown to bits. Equipment was wrecked; Jap bodies and body parts were everywhere. They said over a thousand dead Japs were in the riverbed, but a real body count wasn't possible. A few days later a big storm flooded the riverbed and washed the bodies out to sea."

"At night the Japs would crawl up to our perimeter looking for a weak spot. Anything that moved at night we considered to

be the enemy and to be killed. One night a Jap got inside the perimeter and killed some of our guys while they slept. He was killed when discovered but the damage had been done. Nobody slept for days after that. He got inside the perimeter because a guard fell asleep. I heard later that he was sent to Federal Prison in Levenworth, Kansas. The Japs would call out from the jungle, 'Soldier, you die tomorrow', or 'American, you die'. You could hear the Japs out in the jungle getting "boozed up" before a Banzai Attack. After such an attack bulldozers moved in and dug trenches and pushed their rotting bodies in and covered them up."

"A few of the guys were antsy and went out on their own patrols for something to do. They said they were going out to look for Japs that got through and kill 'em. Some guys said they weren't supposed to do that without orders, but another soldier said let 'em go."

The Philippines

"Some of the guys were still with the old company, but by the time we got to the Philippines in 1944 a lot of us had been transferred or re-assigned. I was Supply Sergeant in the 792 Base Depot Company and Transportation Unit. I took part in combat operations in the liberation of Leyte in the Southern Philippines. I had one luxury as Supply Sergeant; I had a little Filipino girl that did my laundry for me."

Roland was responsible for requisitioning, receiving, storing clothing, supplies, equipment, ammunition and weapons and getting them to the front.

An image of friends, the boys from Sheboygan Falls who stayed together throughout the war. Taken in Australia, April 1943.

Upper left: George Bowser, Upper right: Lowell 'Kid' Lloyd, Lower left: Floyd LeMahieu Lower right: Erwin Heidenreiter. Not pictured: Verlin 'Bud' Norgaard. Bud was killed earlier on January 1, 1943 at Buna.

Above: The first wave of troops going ashore at Aitape, New Guinea.

Below: Moving up the beach to the front at Aitape.

"When MacArthur came ashore a landing barge came up on the beach and some officers got out and looked around. They talked a bit and then got back in the barge. The barge pulled a little ways off the beach and dropped the ramp. MacArthur made his dramatic entrance down the ramp into the water. The photographers and newspaper people recorded his triumphant return to the Philippines. It was just a big show."

"Vic Felde was now with the 63rd Infantry 6th Infantry Division. I later learned he had been killed somewhere on Luzon on January 17, 1945. He had made it through New Guinea only to die so close to the end."

Newsreels captured MacArthur wading ashore at Leyte returning to the Philippines as he had famously promised he would after the original defeat of American forces there. What the newsreels didn't capture were the 67 days it took to subdue the island, with the loss of more than 55,000 Japanese soldiers during the two months of battle and approximately 25,000 more soldiers killed in smaller-scale engagements necessary to fully clear the area of enemy troops. The U.S. forces lost about 3,500 soldiers. (History.com)

Home Again

Dad had accrued more than enough combat /service points to be rotated out. He had served 1 year, 8 months and 20 days in states, and 3 years, 1 month and 10 days in the Pacific. He had lost some good friends. These were guys he grew up with. They trained together, drank beer together, played ball together and just plain had some good times together. They had a bond few ever know much less understand. They were young and cocky and thought they were ready for anything.

Things had changed a lot since Dad left home in 1940 and he wondered how it would be. When he passed under the Golden Gate Bridge he said he knew that he had finally come home. He arrived in Sheboygan in June 1945 and on June 30, 1945 he married Doris Leonhardt at St. Dominic's Catholic Church in Sheboygan, Wisconsin.

Dad was honorably discharged on August 14, 1945 at Fort Sheridan, Illinois. Roland and Doris had four children. Me, Richard Allen: born June 22, 1946, my sisters; Linda Lou born April 13, 1948, Patricia Ann born October 30, 1949 and my brother Daniel Jay born November 30, 1951.

Roland Stoelb and Doris Leonhardt in Sheboygan on Roland's first day home, June 1945.

June 30, 1945 - Wedding Day for Roland A. Stoelb and Doris E. Leonhardt. Front row, left to right- Beverly Christenson, bridesmaid, friend of the bride; Marion Worscheck, bridesmaid, friend of the bride; Sidonia Saaman, bridesmaid, sister of the groom; Joyce Ann Merrill, bridesmaid, cousin of the bride; Blossom Zastrow, Matron of honor, sister of the bride; Doris Leonhardt, bride; Roland Stoelb, groom; Judy Kleist, flower girl, niece of the groom; Jerry Reinke, ring bearer; Back row, left to right- Sgt. Francis Hauch, groomsman, friend of the groom, brother of Lefty; S/Sgt Steve Engl, groomsman, friend of the groom; Sgt. Hilarian (Lefty) Hauch, best man, friend of the groom.

Back in the Army and Korea

After the war, and for some years, there was a housing shortage. With the massive number of soldiers, sailors and marines coming home the returning veterans found it difficult, if not impossible, to find a place to live. My dad was no different in that regard. From 1945 to 1948 Mom and Dad lived with Mom's parents Harold and Lucille Leonhardt at 2404 N 6th St (6th and Bell Av) in Sheboygan. The United States was still in transition from a war time economy to a peace time one and many domestic items, and jobs, were still hard to come by. So, married and with two children to support Dad re-enlisted in the United States Army on July 9, 1948. I have no idea how he passed the physical; but, he did. In the Army he was back in supply, a job he knew well. Besides receiving good pay he also received housing and medical benefits for his family.

Roland and Doris had four children. Richard Allen (b. June 22, 1946), Linda Lou (b. April 13, 1948), Patricia Ann (b. October 30, 1949) and Daniel Jay (b. November 30, 1951). My sister Linda and I were born in Sheboygan, Wisconsin and Pat and Dan were born at Fort Riley Kansas in the post hospital.

On January 7, 1950 at Fort Riley Kansas he re-enlisted for an "indefinite period". The Korean War began in June 1950 and eventually Dad was sent to Korea. He spent 16 months overseas being awarded two bronze battle stars. Dad talked about how cold winter in Korea was. Sitting in the truck with the heater on his feet still got frostbite. Dad saw some of our boys being released by the North Koreans. Some of them refused to come back and elected to stay in North Korea. He saw little kids digging in the garbage cans outside the mess hall picking out food we had thrown away. When I heard that I understood what he meant when we refused to eat something on the table and he would say; "Eat it. You're darn lucky to have that."

Under the terms of his "indefinite period of enlistment" he resigned from the Army and applied for disability. He was discharged in July 1954 at Fort Gordon Georgia.

Epilogue

When I was a young boy I remember seeing Dad wrapped up sweating and shaking. I asked Mom "What's wrong with Daddy?" she just said he had the flu. It was years later that I realized he still suffered from the effects of malaria. His feet also bore testimony to his ordeal. His toe nails were permanently discolored, deformed and thickened due to contracting a jungle fungus infection. Dad took Lithium, a mood stabilizer for "neurosis, depression and anxiety reaction". Today we call it Post Traumatic Stress Disorder (PTSD).

A Japanese grenade killed a man next to him on New Guinea and injured Dad's right ear drum. He recovered from the concussion and his hearing returned, but the damage had been done, and over the years the hearing in that ear gradually deteriorated. He applied for the Purple Heart, but the Army denied the application due to the lack of sustaining evidence; the injury had not been treated by medical personnel, and medical treatment was not part of official records.

I wonder what he thought every Thanksgiving Day and especially Christmas Day. I don't think a Christmas Day went by where sometime that day he didn't think of that "Green Hell Hole" of New Guinea and his buddies. And like any combat veteran who lost friends he felt both gratitude and guilt and asked himself "*Why them and not me?*" It is one of those unanswerable questions. However, when he came home in June of 1945 it was a time to celebrate and live his life for those that had given theirs.

In 1975, at the age of 55 he was considered unemployable by the Veterans Administration due to his service connected disabilities of heart disease, anxiety reaction and defective hearing. He was awarded increased compensation due to the increased severity of his physical condition. Even with this recognition by the VA of his defective hearing as being service connected he still didn't get the Purple Heart awarded retroactively.

In 1960, I remember when Dad went to the VA Hospital in Milwaukee to have his gall bladder removed. One time I went with Mom to see him. She said that sometimes she would have to wait for him because he would be upstairs visiting some of his buddies; men who had lost both their arms and legs and men badly disfigured in war. Some of these men refused to go home because they didn't want their families to see what war had done to them. But, Dad went to them. This was the kind of man my dad was and I am so extremely proud of him.

He worked as a carpenter, a mason, a mechanic, ran the 99 Hall in Sheboygan, and was a machinist and a school bus driver. No matter what life through at him he maintained his dignity and remained a proud and honest man until his death. Dad died on January 2, 1987. He was 66 years old, a death that was deemed to be 100% service connected.

I read the following quote somewhere and thought it to be a very fitting ending to this story.

Real heroes don't wear capes; they wear dog tags.

There is nothing more to say.

Dad, Mom, Dickie and little sister, Linda in 1948 just before Dad went back into the Army.

Dickie and Linda, Fort Riley, Kansas, 1952.

Patty, Linda and Danny, May, 1954 at Grandma Leonhardt's house in Sheboygan.

Dad , Mom and Dickie, May 1954. Dickie's First Communion, Sheboygan. Dad had just returned from Korea. After 30 days leave he drove the old Ford to Georgia for discharge in July.

Awards and Decorations received by Roland A. Stoelb

World War II Pacific Theater

Good Conduct Medal

Distinguished Unit Badge

American Defense Service Medal

American Campaign Medal

World War II Victory Medal

Asiatic Pacific Campaign Medal w/2 bronze battle stars

(Papuan Campaign: Buna-Sananada on New Guinea)

(Aitape: Northern New Guinea)

Philippine Liberation Ribbon w/1 bronze battle star (Leyte)

6 Overseas Service Bars and 1 Service Stripe

Combat Infantry Badge: awarded for being personally present under hostile fire, while serving in assigned, primary infantry or Special Forces duty in a unit actively engaging the enemy in ground combat.

Bronze Star: established and first awarded February 4, 1944 but retroactive through December 7, 1941; awarded for bravery, acts of merit or meritorious service: Mom said he went out under fire to retrieve a buddy from the battlefield.

Korean War

National Defense Medal,

United Nations Service Medal

Korean Service Medal with 2 bronze battle stars.

***A bronze battle star or campaign star indicates participation in a battle or campaign.**

Sheboygan Rosters from October 1940
Company F, 127th Infantry
Sheboygan Wisconsin

KIA: Killed in Action
WIA: Wounded in Action

Officers:
Captain Ernst Pelchen, Jr.,

1st Lt. Henry W. Fahres,

2nd Lt. Marcellus P. Fahres,

2nd Lt. Alfred E. Meyer (WIA; Dec 1942-Buna, New Guinea)

2nd Lt. Elmer B. Ehren (WIA; Dec.18, 1942 - Buna, New Guinea)

Non-Commissioned Officers:
First Sgt: Peter P. Heymanns

Sergeants:
James D. DeMets

Frederick Dottai

Henry Dotz (WIA; Dec 1942-Buna, New Guinea)

Frederick H. Fetterer

Lyle R. Perry

Francis Schommer

Martin J. Toebak

Corporals:
Christ M. Brock, Jr

Warren T. Dean

David Dottai, Jr

Ira Giefer

Waldemar M. Hirt

Lester J. Holst; (KIA; Dec 1942-Buna, New Guinea)

Frederick Jansohn

Donald Likas

Roger J. Ross

John C. Schoemer

Privates First Class:

Herman W. Gesch

Harold A. Heck

Christ Knaub

Carl F. Knocke

Harry A. Nischik

Warren E. Schneider; (KIA; Jan 1, 1943-Buna, New Guinea)

Earl Skelton, Jr.

Privates:

Harly J. Back

George J. Bowser (WIA; Dec. 24, 1942-Buna, New Guinea)

Roman N. Bowser

Elmer Brock

Jerome R. Cudworth

Christ J. Dahmer (WIA; January 8, 1943 - New Guinea)

Frederick A. Eisner

Reuben T. Engelhardt

Steve A. Engl (WIA: December 1942- Buna, New Guinea)

Franklin Fisher

John Felde

Victor Felde; (KIA; Jan 17, 1945-Luzon, Philippines)

Cornell J. Gehl

Francis M. Hauch

Hilarian J. Hauch

Erwin P. Heidenreiter

Victor Henze

John Hermann, (KIA; Dec 25, 1942-Buna, New Guinea)

Fred Niesing, (WIA; Dec 30, 1942-Buna, New Guinea)

Bernard E. Hoehnke

Ernest F. Holtz

Edward A. Karbe

Robert J. Keitel; (KIA; Dec 24, 1942-Buna, New Guinea)

Norbert P. Koene

Robert Kohn

Joseph Kolenc;

Arthur G. Kroos, Jr.

Robert R. Krueger

Carl H. Lange

Floyd Le Mahieu; (WIA; Dec 30, 1942-Buna, New Guinea)

Lowell R. Lloyd

Frederick H. Matthias

Anton (Tony) Mesich

Verlin S. Norgaard; (KIA; Jan 1, 1943-Buna, New Guinea)

Robert J. Oostdyke

Donald A. Pauly

Gordon J. Pichl; (WIA; Dec 28, 1942-Buna, New Guinea)

Donald C. Rothe; (WIA; Dec 31, 1942-Buna, New Guinea)

Roman F. Richter;

Kurt G. Schweingruber

Frederick G. Sperb

Bernard R. Stegemeyer; (KIA; Dec 25, 1942-Buna, New Guinea)

Frederick Stoehr

Roland A. Stoelb

Charles A. Strobel

Roland J. Tilton

Lester Vanderweele; (KIA; Dec 25, 1942-Buna, New Guinea)

Walter S. Verstrate

Arthur H. Weber

Carl P. Weiss

Frederick H. Wesener; (WIA: Jan. 8, 1943 - Sananda, Buna Campaign, New Guinea)

Wesley R. Wilson

Service Battery, 120th Field Artillery, Sheboygan, Wisconsin

Officers:

Captain Emil F. Stielow,

1st Lt. Erwin P. Wolf

2nd Lt. Henry W. Honold

2nd Lt. Edward J. Nyenhuis

Master Sergeants:

Frank F. Royce

Gilbert C. Wilke

Staff Sergeants:

Ambrose J. Rewolinski

Edward R. Wirt.

Sergeants:

Peter Fields

Erwin P. Wolf.

Corporals:

Carl A. Brickner

Privates First Class:
Elsworth Anderson

Victor Guenther

George A. Haen

Ewald Kuhlman

Robert G. Nick

Reinhold W. Stichert

Paul H. Troster

Carlton J. Wolf

Werner Wolf

<u>Privates</u>:

John E. Bainbridge; (KIA; Dec 1, 1942-Buna, New Guinea)

Lester Berger

Lloyd O. Boyce

Warren Friederichs

Rudie A. Glazer

Arthur J. Hinkel

Joseph J. Jerale

August A. Krenn

Arthur Maertz

Walter Markgraf

Arthur Marquardt

Harold C. Miller

Henry A. Mueller

Edward L. Nelson (Killed; Dec 2, 1940-Alexandria LA)

Norman E. Ringle

Christian Seng

Harley G. Steuber

Edgar H. Thimming

Clarence Yurk

Leroy H. Zimmermann

Charles W. Berger

John Brack

Lester Damrow

Arthur N. Dern

Edward Diener

Fred Hayes

John Helzsart, Jr.

Robert J. Heilberger

Frederick Krimel

Merlyn H. Reimer

Charles Skoris

Louis L. Stubler

Medical Department, 107th Q.M. Regiment, Sheboygan Wisconsin

Officers:
Major Harry H. Heiden

Capt. Wylie James Ferris

Non-Commissioned Officers:
First Sergeant: Frederick Raih

Sergeants: Ernst Brack; Henry Schaefer

Privates First Class:

Christ Felde

Carl C. Diener

Privates:
Albert Becker

Kenneth Krueger

Ervin W. Mueller

Anthony Price

Ray A. Scherhaufer (KIA; Mar 9, 1943-SW Pacific)

Leonard Trass

A number of men from Sheboygan and the surrounding area joined the 127th after Pearl Harbor and also fought on New Guinea. It is not my intention to leave anyone out; however, this story deals with the men of the Sheboygan Wisconsin National Guard Unit listed on the rosters as of October 15, 1940.

32nd Division, 127th Infantry, Wisconsin		
Company	**Battalion**	**City**
A	1st	Marinette
B	1st	Rhinelander
C	1st	Oconto
D	1st	Appleton
E	2nd	Manitowoc
F	**2nd**	**Sheboygan**
G	2nd	Oconomowoc
H	2nd	Oshkosh
I	2nd	Neenah
K	2nd	Milwaukee
L	2nd	Jefferson
M	2nd	Fond du Lac

32nd Division, 128th Infantry, Wisconsin		
Company	**Battalion**	**City**
A	1st	Menomonie
B	1st	Eau Claire
C	1st	Marshfield
D	1st	Rice Lake
E	1st	Beaver Dam
F	1st	Portage
G	1st	Madison
H	1st	Baraboo
I	1st	Stoughton
K	1st	Monroe
L	1st	Beloit
M	1st	Platteville

Sources and References

Victory in Papua, The War in the Pacific (Office of the Chief of Military History, Dept. of the Army, Series U. S. Army in-World War II: by Dr. Samuel Milner (1957)

The 32nd "Infantry Division in World War II: by Maj. Gen. H. W. Blakeley USA, Ret. (1955)

Bloody Buna by Lida Mayo (1974)

The Ghost Mountain Boys by James Campbell (2007)

I was an Infantryman in World War II by Floyd LeMahieu (Sheboygan Falls, Wisconsin) Company F, 127th Infantry, 32nd Division (1990)

Sheboygan Press

Milwaukee Sentinel

Roland A. Stoelb (Sheboygan, Wisconsin) Company F, 127th Infantry, 32nd Division

Crusade in Europe: Dwight D. Eisenhower, Doubleday and Company, Inc. 1948

U.S. Army Signal Corps, public domain images

The Ultimate Sacrifice by Elmer Koppelmann, 2004

Department of the Navy - Naval Research Center, Washington D.C.

Made in the USA
Monee, IL
04 November 2026

67382708R10059